HERB LIBRARY

MINT

Kate Ferry-Swainson has been a writer and editor for fifteen years, specializing in gardening, history, mythology, and self help. As a writer she has most recently contributed extensively to *Mindpower*, a major series of self-help books; has co-written *Spirit Stones*, a book about Native American spirituality, and has written the historical section for the art historical monograph *Van Dyck*. She has also written *Herb Library: Camomile* and *Herb Library: Ginger* in the same series as this volume.

Deni Bown is a freelance writer and photographer specializing in botany, gardening, herbs, and natural history. The author and photographer of *The RHS Encyclopedia of Herbs and Their Use*, her other books include *Growing Herbs* and *RHS Plant Guides: Garden Herbs*. As President of The Herb Society, she lectures on various aspects of herbs and other plant topics, and has travelled throughout the world in pursuit of her subject as well as conducting specialist garden tours. She is a regular researcher and photographer at both Kew and Edinburgh Botanical Gardens.

This book gives general information about how to use herbs and essential oils. The material in this book is not intended to take the place of the diagnosis of a qualified medical practitioner. The author, editor and publisher cannot accept responsibility for any side effects caused by the use or reliance on any of the information herein.

Dedication:
For Richard, Matilda and Wriggly Baby

THIS IS A CARLTON BOOK

This edition published by
Carlton Books Limited 1999
20 Mortimer Street
London
W1N 7RD
www.carlton.com

A CIP catalogue for this book is available from the British Library.

ISBN 1 85868 697 0

Editorial manager: Penny Simpson
Design: blu inc.
Picture research: Frances Vargo
Production: Alexia Turner/Garry Lewis
Printed in Dubai

HERB LIBRARY

MINT

Kate Ferry-Swainson

Series Editor: Deni Bown

CARLTON

CONTENTS

INTRODUCTION

If any man can name the full list of all the kinds and all the properties of mint, he must know how many fish swim in the Indian Ocean.
Wilafrid of Strabo, **AD875**

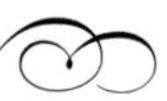

Mint is a prolific, highly aromatic plant that has been used in many different ways by cultures as diverse as the ancient Chinese and the Native Americans. Civilizations all over the world and throughout history have valued it for its dual qualities: to cool down and to warm up. It is a herb that has been used in many ways: medicinally, to aid digestion, decongest a chest and ease nausea; in cosmetics to clean oily hair, freshen the breath and provide a cooling massage; and in homes to repel insects or rodents and to create a cool, breezy atmosphere.

This book shows you how to identify the main therapeutic species and cultivars of mint, and then how to use them to make your own herbal remedies, cosmetics and scented gifts. The invaluable role that mint has played throughout history – entering the fabric of a society via its mythology and spiritual belief system – is also discussed.

VARIETIES OF MINT

There are about twenty-five different species and various hybrids and cultivars of mint from the family *Labiatae/Lamiaceae*. It is difficult to count the exact number since both in the wild and in cultivation they cross-pollinate easily and hybrids occur. The genus *Mentha* consists mainly of perennials, with a few annuals, which flower in summer or early autumn. The plants are native to temperate regions of Europe, Asia and Africa. Crops for both leaves and essential oil are now grown on a large scale in Europe, the USA, the Middle East and Asia.

Mint grows in the wild in wet places such as a stream.

There is a wide range of leaf colour and shape, stem and flower colour and habit.

PEPPERMINT, MENTHA X PIPERITA

One of the most fortuitous of hybrids is the cross between watermint, *Mentha aquatica*, and spearmint, *Mentha spicata*, which produced peppermint, *Mentha* x *piperita*, perhaps the most celebrated and most popular species of mint used in herbal medicine today.

The fresh, rousing greenness of peppermint.

Peppermint is very popular because of its very high menthol content, which means that it is the most rousing of all the mints. However, the high concentration of menthol means that peppermint can be an irritant. Menthol is both cooling and warming and acts as an antiseptic, a decongestant, an analgesic and an anaesthetic. It is extremely valuable as a medicinal substance.

The peppermint plant is a creeping perennial with smooth leaves and lilac-pink flowers. Its stems may be tinged with purple. It exudes the distinctive menthol smell. It is non-toxic but may cause irritation, especially to children.

Medicinally, peppermint is used internally for 'flu, colds, colic, irritable bowel syndrome, indigestion, gastric ulcer, any kind of nausea, including sickness in pregnancy, diarrhoea and sore throat. Its external applications include treatments for sinusitis, asthma, catarrh, rheumatism, itchy skin, burns and neuralgia. It also acts as an insect and small-mammal repellant.

The cultivar, *M.* x *piperita* f. *citrata* – eau-de-Cologne mint or lemon mint – is popular for its scent, which is like lavender. This is used for nervous exhaustion and infertility.

Right: Peppermint is used in remedies to treat a wide range of common complaints.

SPEARMINT, MENTHA SPICATA (SYN. M. VIRIDIS)

Spearmint has a sweetish aroma and, in this case, lilac flowers.

Spearmint has a lower menthol content than peppermint; this makes it less of an irritant and therefore more suitable for use with children. Indeed, it is considered an adequate substitute for peppermint in many remedies for children.

The spearmint plant is a creeping perennial with wrinkled leaves and lilac, pink or white flowers. It is more sweetly scented than peppermint. Spearmint is non-toxic, non-irritant and non-sensitizing.

This species is used in the treatment of indigestion, colic, feverish illnesses and many childhood ailments. Spearmint is very popular in oral hygiene preparations.

PENNYROYAL, MENTHA PULEGIUM

Pennyroyal is a small, creeping perennial with lilac flowers and a peppermint-like aroma.

Pulegone, a toxin that stimulates the uterus and may cause abortion, is found in this species, and it is therefore not given to pregnant women. Medicinally, pennyroyal is used for period problems, colic, indigestion and feverish colds; it is also used externally for skin irritations. Pennyroyal volatile oil is an oral toxin so care must be taken not to ingest this species in large doses, as this has resulted in death. It repels rats, mice and insects.

Pennyroyal has a fresh aroma and lilac flowers.

WATERMINT, MENTHA AQUATICA

Watermint, as its name suggests, grows by rivers and streams.

Watermint is a spreading perennial with purplish-red stems, hairy leaves and lilac flowers. It has a strong scent reminiscent of peppermint or pennyroyal.

It is used medicinally for indigestion, spasms, diarrhoea, colds and painful periods.

CORN MINT, MENTHA ARVENSIS

Corn mint, or field mint, is a spreading annual or perennial plant with hairy leaves and lilac-pink flowers. Its leaves have a rather bitter smell. It is non-toxic, but may cause irritation.

It is used medicinally to aid digestion, ease spasms, calm itching and reduce inflammation. It also suppresses milk production in women, so breastfeeding women beware. This species plays a valuable role in traditional Chinese medicine, where it is known as *Bo He*.

Two varieties of this species are *M. a.* var *piperascens*, Japanese corn mint, and *M. a.* var. *villosa*, American mint. Japanese corn mint is South-East Asia's predominant source of menthol. American mint was known to the Native Americans.

Right: Corn mint flourishes in the wild, and is valuable in cultivation.

Left: Applemint is valued for its gentle, apple-like aroma.

OTHER MINTS

Applemint (*Mentha suaveolens*) is a well-known type of mint used for cooking and in cosmetic and scented decorations. It has a gentle, sweet scent with mellow overtones of apple. It is a creeping plant with wrinkled leaves and pink or pink-white flowers. Pineapple mint (*Mentha suaveolens* 'Variegata') is a variegated relative of applemint.

Gingermint (*Mentha* x *gracilis*) is a hybrid between spearmint and corn mint. It has a spicy, peppery flavour valued in cooking recipes, and combines particularly well with melon, other soft fruits, and tomatoes. The stems of gingermint are tinged with red, the leaves are smooth, and the plant produces lilac flowers.

Bowles' mint (*Mentha* x *villosa* var. *alopecuroides*) is a hybrid between spearmint and applemint. It has slightly hairy leaves and purple-pink flowers. It is used to make mint sauce or jelly, and combines well with new potatoes and peas.

Gingermint is a valuable ingredient in the kitchen.

Mint makes an attractive addition to a mixed border.

CULTIVATING MINT

Mints are hardy plants: that is, they survive outside during the winter months in a temperate climate. They like to be planted in rich, moist soil, in sun or partial shade. Mints grow well in a herb garden, container or mixed border.

All mints are spreading and, left to their own devices, will quickly overrun their allotted space. If planting them in a border, it is a sensible precaution to grow them in a somewhat confined space to restrict their growth. This may easily be achieved by planting mint in a container (a plastic pot or bucket) slightly larger than the plant is at present. Make drainage holes in the bottom of the pot and fill it with a mixture of soil and compost. Then sink the container into the allotted space. At the start of each spring dig up the container, divide the rhizomes, discard any excess, and replace the soil and compost mix.

ROUTINE MAINTENANCE

There is little routine maintenance involved with growing mint. Water the plants thoroughly immediately after planting them, every day for the first week and thereafter in dry periods.

Remove weeds regularly to prevent the plants from becoming choked. Invasive herbs such as mint need to have their growth limited on a regular basis. Mints produce runners which will quickly spread. Cut these off as you find them.

Mint produces bright new, juicy foliage if you cut it right back a short while before it flowers. In this way you can harvest the young leaves. Harvesting large quantities of mint is an effective way of restricting the growth of the plant.

Plants in containers need to be looked after with a little more care in times of heat and cold: in dry periods, water containers such as strawberry planters well, and protect them from frost during winters.

HARVESTING AND STORING MINT LEAVES

In previous centuries much ritual and hype developed surrounding when to harvest various herbs in order to ensure that they were picked in peak condition. Nicholas Culpeper, (1616–54), who wrote a definitive herbal, which was published in 1653 and which is still in print to this day, was rather keen on astrology. He said the following:

> *Let the planet that governs the herb be angular, and the stronger the better; if they can, in herbs of Saturn, let Saturn be in the ascendant; in the herbs of Mars, let Mars be in the mid heaven, for in those houses they delight; let the moon apply to them by good aspect, and let her not be in the houses of her enemies; if you cannot well stay till she apply to them, let her apply to a planet of the same triplicity; if you cannot wait for that time neither, let her be with a fixed star of their nature.*

In essence, however, the advice is straightforward enough: choose a dry day to harvest the whole plant or simply the leaves just before the plant flowers. Make sure the leaves are clean, undamaged and uninfested by insect pests. If harvesting the whole plant pick it over to remove any unsuitable leaves. Do not wash the leaves or rub them clean. Handle them as little as possible: the more you touch the leaves the more essential oil rubs off on your fingers, making the leaves less effective in whatever preparation you intend them for.

Culpeper advises, 'Of leaves, choose such only as are green, and full of juice; pick them carefully, and cast away such as are any way declining, for they will putrify the rest. So shall one handful be worth ten of those you buy in any of the shops.'

Pick just as many leaves as you need: enough to make a day's supply of tisane, if appropriate, or enough to make a jarful of tincture or ointment. If necessary, fresh leaves may be stored for a short time, packed in plastic bags in the fridge.

If harvesting whole plants, or stems, hang them upside down in bunches in a warm, dry and well-ventilated area such as an airing cupboard. The quicker they dry, the more effective they will be in your preparations because the more of their aromatic qualities will be retained. If you choose to harvest leaves only, place them on a clean and dry tray in a warm, dry place. When the leaves are brittle, rub them gently onto a piece of clean paper, then make a chute out of the paper and funnel them into a dark glass jar.

BUYING MINT

If you are buying mint – whether in the form of whole plants or fresh or dried leaves – it is imperative to make sure that you know exactly which species you are purchasing. Since mint hybridizes so freely in the wild, making it difficult always to identify species with certainty, ensure that you buy plants and leaves from established outlets, including mail-order companies. Do not pick mint from the wild. If you are picking from a plant pre-existing in your garden seek botanical advice to identify its species.

Chinese herbal medicines on display in a Hong Kong pharmacy

When buying whole plants, check them over thoroughly to ensure that they are free from pests and disease, have a healthy root system with strong roots well distributed throughout the compost, and are not pot-bound: that is where the plant has outgrown the pot and the root system has no space and is running round and round the base of the pot or even escaping out of the bottom.

USING HERBAL REMEDIES

Although the kinds of mint discussed in this book are safe, as detailed, for use in medical remedies and cosmetic preparations, herbs should be used in moderation. In particular, mint should be drunk in the form of tisanes no more than three times a day. Especial care should be taken with children and if you are pregnant or breastfeeding. Peppermint can cause an allergic reaction because its menthol content is extremely high. Particular care should be taken when using mint essential oils as they can irritate the mucous membranes. Taken in excess, all herbs become toxic; used sensibly following the guidelines given in the following pages, mint may be used safely and effectively.

Minor, everyday illnesses can be treated very effectively using herbal remedies. However, great care should be taken in pregnancy and with babies and elderly people. Please note that peppermint should not be given in any form to babies or children of less than four years as it can irritate the lining of delicate stomachs. Instead use spearmint, or catmint (also known as catnip) – *Nepeta cataria* – which is a gentle herb that contains no menthol but which is an effective substitute in treating some children's ailments (see page 57).

Be honest with yourself when analysing your symptoms and the possible causes of your ailment. Herbal remedies are effective treatments for short-term complaints, but should not be used to treat ailments caused by your way of life, such as obesity, stress or depression. If you think that this applies to you, try to change your lifestyle instead. Do consult a medical, herbal or homeopathic practitioner if you have a more serious complaint, if your ailment does not improve with herbal remedies, or if you wish to seek individual advice about your particular symptoms.

If you are already taking some form of medication, seek professional advice before taking herbal remedies. In particular, do not mix mint herbal remedies with homeopathic preparations as mint essential oils cancel the effect of homeopathic remedies.

This book gives general information about how to use herbs to make herbal remedies and cosmetics. The author, editor and publisher cannot accept responsibility for side-effects caused by taking the herbal remedies discussed in this book.

Planted in an old bucket, mint thrives but does not become rampant.

CHAPTER 1

HISTORY

AND, BEHOLD, A WOMAN IN THE
CITY, WHICH WAS A SINNER, WHEN
SHE KNEW THAT JESUS SAT AT MEAT
IN THE PHARISEE'S HOUSE, BROUGHT
AN ALABASTER BOX OF OINTMENT, AND
STOOD AT HIS FEET BEHIND HIM WEEPING,
AND BEGAN TO WASH HIS FEET WITH
TEARS, AND DID WIPE THEM WITH THE
HAIRS OF HER HEAD, AND KISSED HIS FEET,
AND ANOINTED THEM WITH THE OINTMENT.

LUKE, VII:37–8

Mint was used as a remedy in China up to 4,000 years ago. The countries around the eastern Mediterranean also discovered the aromatic and therapeutic qualities of the herb and used it in many ways, not least of which was for its efficacy as a soothing, cooling and relaxing treatment for hot, tired feet. Indeed mint became part of the ancient Greek fabric of life and belief, to the point where a myth was established about the origins of the plant. The Arabs clung on to ancient Greek scholarship at a time when Europe was sliding into the dark ages, and the ancient learning and practices were re-introduced into Europe via the Moors.

This illustration from the Latin manuscript *The Art of the Apothecary* shows a collection of plants and making of medicines.

This ancient learning was embraced warmly by medieval apothecaries who found that it complemented their own traditional understanding of how to treat various diseases. It was in the Renaissance that further advances were made in the understanding of how the human body worked, how to diagnose and treat disease, and, together with the invention of the printing press, a new breed of medical and herbal texts was published. These were in use right up until the scientific revolution of the last century, and today we see a renewed interest in using plants and herbs to treat disease.

CHINA AND SOUTH-EAST ASIA

Chinese medicine is widely considered to be the oldest surviving system of herbalism in the entire world, with an unbroken history stretching back over 4,000 years. In Chinese medicine *Mentha arvensis*, corn mint, is a valuable substance known as *Bo He*, used to treat earache, skin problems and tumours. In the West, *bohea* has been a term for China tea. In eighteenth-century polite and fashionable society *bohea* was widely drunk. It is now known as a low-quality, black, China tea.

The basic system of Chinese medicine was established 3,000 years ago by Huang Di, the first Emperor of the Han dynasty, the legendary Yellow Emperor. He documented the system of Chinese medicine in a book on the subject.

The gods of Chinese medicine: Huang Di, Fu Shi and Shen Nong.

This basic system is still followed and holds that nature is composed of five elements: wood, fire, earth, metal and water. The elements are responsible for our well-being and for the balance that is needed in our bodies and in nature in order to achieve harmony and health. Illness is caused by six evils: wind, heat, cold, dryness, dampness and summer heat.

Each element is believed to interact with the others and to be linked to one or more organs, emotions, tastes and seasons. It is believed that specific herbs strengthen different elements in the body: *Bo He* acts on lung and liver energies.

Several ancient Chinese herbals are still in regular use today. The first major one was *The Yellow Emperor's Classic of Internal Medicine*, written in about 1,000BC by a number of authors and attributed to the Yellow Emperor himself, Huang Di. Another book, called the *Shen Nong Canon of Herbs,* lists 252 drugs made from herbs. Shen Nong was an emperor-god who lived in about 3,000BC; the book was written in his name by a collection of people between AD25 and 220. Both of these pioneers were Taoists, and their herbals abound with Taoist spiritual wisdom. They advocated that people should embrace virtue in order to live long and to prosper. Herbs would help them in their striving to live virtuously.

A miniature of the emperor-god Shen Nong.

The theory of *yin* and *yang* is important in Chinese medicine: opposites needing to be in harmony to achieve health and well-being. According to Nei Jing, writing in the first century BC, the Yellow Emperor said 'Ying/Yang are the way of Heaven and Earth, the great principle and outline of everything, the parents of change, the root and source of life and death, the palace of gods. Treatment of disease should be based upon [their] roots.'

A Chinese herbal medicine store.

In the early twentieth century, Chinese medicine started to fall out of favour, in much the same way as traditional herbalism did in the West, as modern medicine began to be seen as the superior system. However, in China and in the West, the 1960s saw a revival of interest in traditional Chinese medicine, and *The Atlas of Commonly Used Chinese Traditional Drugs* was compiled and published by the Chinese Academy of Sciences in 1970.

Chinese traditional medicine had an important influence on the development of herbal medicine in the West. Beliefs about the origins of disease, as well as knowledge of plants and recipes for treatment, were carried westwards along the silk and spice routes. Elsewhere in Asia, mint became an important part of everyday life. In Japan, the nobility wore pomanders of mint suspended from sashes on their kimonos; as they walked the aroma wafted around them and helped them to stay alert.

ANCIENT GREECE

The ancient Greeks developed a great tradition of using mint in remedies. In Athens, there was a fashion for using mint oil to massage the arms for a refreshing effect. Spearmint was used to scent bath water and to restore flagging energy. At this time medical remedies embraced what we now consider to be cosmetic treatments. Advances made by the great physicians of this time were pre-eminent in Europe until the Renaissance.

HIPPOCRATES

Hippocrates, the father of medicine in the West, who lived from 460 to 377BC, firmly set the direction that Western medicine was to take, in both the understanding of the causes of illness and its treatment. There are remarkable similarities between the system of Hippocrates and that developed in traditional Chinese medicine. Hippocrates believed that everything in nature is made up of elements, namely earth, air, fire and water, and that man is intimately connected to these elements. For the seasons to ebb and flow and for life to take its natural course, these elements had to be in harmony. The four elements governed the four humours that resided in the human body: phlegm, blood, black bile and yellow bile. Health depended on these four humours being in proper balance.

Unlike the five elements in Chinese medicine, Hippocrates' four humours were distinct and did not interact. The humours

Hippocrates and Galen pictured together in a thirteenth-century fresco in Anagni Cathedral, Italy.

were thought to affect both physical and emotional conditions: for example, depression and unhappiness were seen as being caused by an excess of black bile.

Hipprocrates took a holistic view of diagnosis and treatment, and believed in curing the person, not the disease. In this respect he analysed not just the body part that was hurting but also the way the person lived, dreamed, exercised and ate. He believed that environment played a large part in the onset of ill-health, and therefore in the re-establishment of health, and planned an individual programme considering herbal remedies, diet, exercise and other issues concerning the individual's environment.
The major writings ascribed to him, the *Hippocratic Corpus*, is a collection of seventy works.

GALEN

Galen of Pergamum, c. AD130–201, ranks second in importance to Hippocrates for the relevance of his work and the endurance of his legacy. He studied medicine in Pergamum, Corinth and Alexandria, and subsequently lived in Rome. He wrote extensively on medicine, collating all the medical knowledge of the time, and became the leading authority for subsequent Greek and Roman medical writers. His writings were taken up by the Arabs who, through the Moors in Spain, transmitted Galen's work to Europe and the West. Indeed, medicine in Europe of the Middle Ages hung on his every word.

DIOSCORIDES

Pedanius Dioscorides (c. AD40–90) was a Greek physician who served in the army of the Emperor Nero, when he made studies of the medicinal properties of plants. His chef d'oeuvre, *De Materia Medica Libri Quinque*, was the first text on botany and pharmacology that was based on science and observation rather than on superstition. He included an account of about 500 plants, including mint, listing their names and therapeutic qualities. This tome was still recognized as the leading authority until well into the seventeenth century.

PLINY

A contemporary of Dioscorides, Gaius Plinius Secundus – Pliny the Elder – was a philosopher rather than a physician. His *Historia naturalis* is a compilation of all the knowledge of the time, rather in the form of an encyclopedia, and mentions all the plants to which he found reference in any book of the time.

THE ROMANS

The Romans made an art of using *aromata*, 'scented things', in medicines. When herbs or spices were first introduced into the culture they were usually far too expensive to be used for food or cosmetics and were reserved for medicinal uses. Spearmint nevertheless became an extremely popular herb for cooking.

Roman women preparing perfumes.

Mint was considered by the Romans – and by plenty of others subequently, including Culpeper – to be an aphrodisiac, and was associated with the goddess of love, Venus. John Keats could be making a reference to mint, among other herbs, when he wrote in his poem *Lamia*:

> *...Where 'gainst a column he leant thoughtfully*
> *At Venus' temple porch, 'mid baskets heap'd*
> *Of amorous herbs and flowers, newly reap'd*
> *Late on that eve ...*

Indeed, the role of aphrodisiacal herbs in love potions can also be seen in the twelfth-century love story of Tristan and Iseult who drank a love potion, *le vin herbé*, and were for ever bound to each other. Ingredients are not given in the legend, but it is quite likely that mint was included. As soon as Tristan and Iseult had drunk the potion, despite themselves they felt bound to each other by the force of their desire.

THE MIDDLE EAST

All over the Middle East and the eastern Mediterranean, references to mint abound. In Egypt a type of peppermint has been found to have existed from tomb paintings dating from 1000BC. Mint was obviously known and used in the lands of the Bible, as the gospel of St Matthew refers to it:

> *Woe unto you, scribes and Pharisees, hypocrites!*
> *For ye pay tithe of mint and anise and cummin, and*
> *have omitted the weightier matters of the law,*
> *judgement, mercy and faith.*
> Matthew, XXIII:23

Indeed, Hebrew temple floors were strewn with mint. Scented oils were also widely used: Mary Magdalen had access to them, presumably making her own massage oil. They were highly prized – particularly if they included spikenard from the Himalayas, which was extremely expensive – as is evident from the apostles' outrage that she should be using it on the feet of Christ rather than selling it for a handsome sum, which could then have been given to the poor.

Yet it is the Arabs whose legacy is undoubtedly the greatest. From the seventh century onwards they preserved many invaluable tomes of Greek medicine, translated them and used them to fire their own scientific revival and the development of their own system of medicine. These tomes would otherwise almost certainly have been lost for ever. Avicenna's *Canon* made a

Left: Armenian ladies of the nineteenth century at home with a tray of mint tea.

synthesis of the best of Greek medical doctrines, including those of Hippocrates and Galen.

Mint tea is a traditional beverage in Arab countries, valued for its cooling properties. Mint has also found its way into many traditional dishes of the Middle East and eastern Mediterranean, including *tzatziki* and *tabbouleh. Tzatziki* is a cooling and fresh dip made with Greek-style yogurt, olive oil, fresh mint, garlic, salt and pepper, traditionally eaten with pitta bread and fresh vegetables. A Middle Eastern dish, *tabbouleh* is a filling and colourful salad made with bulghar wheat, olive oil, lemon, garlic, cucumber, tomatoes, spring onions, parsley, mint, salt and pepper.

NICHOLAS CULPEPER

By the time of the European Renaissance in the fifteenth and sixteenth centuries, the amassed knowledge of Greece, Rome and Arabia, together with the practical know-how of generations of old wives and apothecaries, had been practised throughout Europe for centuries. Men of the Renaissance started to take a real interest in

Nicholas Culpeper, as pictured in his book *The English Physitian.*

discovering how the human body worked, what caused disease, and how to treat it. Even though they might ultimately discard earlier treatises on the composition of the human body (such as in Hippocrates' system) by way of studies on anatomy and dissection, the legacy of the ancients lives on even today. At this time, Elizabethan England, Spain and Portugal were in a great race to conquer the world, to control spice and silk routes, and to expand their geographical and economic boundaries. This widened the horizons of physicians and scientists who set out to discover and catalogue the plant world.

Nicholas Culpeper (1616–54) was a seventeenth-century herbalist whose name continues to have resonance at the millennium. His major work, *The English Physitian: enlarged with 369 medicines made of English Herbs*, referred to as the *Herbal*, is a testament to how much he believed that the individual should be allowed to take part in his or her own cure. The College of Physicians took against him when he translated their *Pharmacopoeia* into English. This enabled ordinary people to read and understand it – and to avoid the need to pay out gross sums to physicians by finding their own herbal medicines in the hedgerows and fields. His own work is a step further in the same direction.

An Elizabethan grocer-druggist shop.

It is interesting to note that in Culpeper's *Herbal*, no mention is made of peppermint, now the most popular variety of herb used in herbal remedies. (Indeed, peppermint was not recognized as distinct from other mints until later in the seventeenth century, and is certainly listed by the time of the

Species Plantarum of 1753.) Instead, Culpeper discusses mint (with particular reference to spearmint) and pennyroyal.

MINT

Of all the kinds of mint, the spear mint, or heart mint being most usual, I shall only describe as follows.

Description. Spear mint hath divers round stalks, and long but narrowish leaves set thereon, of a dark green colour. The flowers stand in spiked heads at the tops of the branches, being of a pale blue colour. The smell or scent thereof is somewhat near unto basil: it increaseth by the root under ground, as all the others do.

Place. It is an usual inhabitant in gardens: and because it seldom giveth any good seed, the defect is recompensed by the plentiful increase of the root, which being once planted in a garden, will hardly be rid out again. Time. It flowereth not until the beginning of August, for the most part.

Government and virtues. It is an herb of Venus. Dioscordes [sic] *saith it hath a heating, binding, and drying quality, and therefore the juice taken in vinegar stayeth bleeding: it stirreth up venery, or bodily lust: two or three branches thereof taken in the juice of four pomegranates, stayeth the hiccough, vomiting, and allayeth the choler. It dissolveth imposthumes, being laid to with barley-meal. It is good to repress the*

Above:
An illustration of spearmint, from *Sowerby's English Botany.*

Left:
The herb garden at Jenkyn Place, Hampshire, showing mint in the foreground.

milk in women's breasts, and for such as have swollen, flagging or great breasts. Applied with salt, it helpeth the biting of a mad dog; with mead and honeyed water, it easeth the pains of the ears; and taketh away the roughness of the tongue, being rubbed thereupon. It suffereth not milk to curdle in the stomach, if the leaves thereof be steeped or boiled in it before you drink it: Briefly, it is very profitable to the stomach. The often use hereof is a very powerful medicine to stay women's courses and the whites. Applied to the forehead and temples, it easeth the pains in the head, and is good to wash the heads of young children therewith, against all manner of breakings out, sores, and scabs therein; and healeth the chops of the fundament. It is also prontable [sic] against the poison of venomous creatures. The distilled water of mint is available to all the purposes aforesaid, yet more weakly. But if a spirit thereof be rightly and chymically drawn, it is much more powerful than the herb itself ... The powder of it being dried and taken after meat, helpeth digestion, and those that are splenetic. Taken with wine it helpeth women in their sore travail in child-bearing. It is good against the gravel and stone in the kidneys, and the stranguary. Being smelled unto it is comfortable for the head and memory. The decoction hereof gargled in the mouth, cureth the gums and mouth that is sore, and mendeth an ill-savoured breath; as also the rue and coriander causeth the palate of the mouth to turn to its place, the decoction being gargled and held in the mouth.

And on the subject of Pennyroyal, Culpeper writes:

Description. The common pennyroyal is so well known that it needeth no description.

There is another kind of pennyroyal, superior to the above, which differeth only in the largeness of the leaves and stalks; in rising higher, and not drooping upon the ground so much. The flowers of which are purple, growing in rundles about the stalk, like the other.

Place. *The first, which is common in gardens, groweth also in many moist and watery places in this kingdom. The second is found wild in Essex, and divers places on the road to London, to Colchester, and parts adjacent.*

Spearmint has been well known since antiquity for its medicinal uses.

Time. They flower in the latter end of summer.

Government and virtues. This herb is under Venus. Dioscorides saith, that pennyroyal maketh tough phlegm thin, warmeth the coldness of any part that it is applied to, and digesteth raw and corrupt matter: being boiled and drunk, it moveth the courses, and expelleth the dead child and afterbirth; being mixed with honey and salt, it voideth phlegm out of the lungs. Drunk with wine, it is of singular service to those who are stung or bit by any venomous beast; applied to the nostrils, with vinegar, it is very reviving to persons fainting and swooning; being dried and burnt, it strengtheneth the gums, and is helpful for those that are troubled with the gout; being applied as a plaister, it taketh away carbuncles and blotches from the face; applied with salt, it helpeth those that are splenetic, or liver-grown. The decoction doth help the itch, if washed therewith; being put into baths for women to sit therein, it helpeth the swelling and hardness of the mother. The green herb bruised and put into vinegar, cleanseth foul ulcers, and taketh away the marks of bruises and blows about the eyes, and all discolouring of the face by fire, and the leprosy, being drunk and outwardly applied: boiled in wine, with honey and salt, it helpeth the tooth-ach. It helpeth the cold griefs of the joints, taking away the pains, and warming the cold parts, being fast bound to the place after bathing or sweating. Pliny addeth, that pennyroyal and mint together help faintings or swoonings, infused in vinegar, and put to the nostrils, or a little thereof put into the mouth. It easeth the headach, and the pains of the breast and belly, stayeth the gnawing of the stomach, and inward pains of the bowels …; it helpeth the falling sickness: put into unwholesome or stinking water that men must drink, as at sea, and where other cannot be had, it maketh it less hurtful. It helpeth cramps or convulsions of the sinews, being applied with honey, salt, and vinegar. It is very effectual for a cough, being boiled in

An abundance of lilac pennyroyal flowers carried on long stalks makes a stunning display.

milk and drank, and for ulcers and sores in the mouth. Matthiolus [Pierandrea Mattioli, see below] *saith, the decoction thereof, being drank, helpeth the jaundice, and all pains of the head and sinews that come of a cold cause; and that it helpeth to clear and quicken the eye-sight. Applied to the nostrils of those that have the falling-sickness, or the lethargy, or put into the mouth, it helpeth them much, being bruised in vinegar, and applied. Mixed with barley meal, it helpeth burnings; and put into the ears, easeth the pains of them.*

Opposite: An illustration of pennyroyal, from *Sowerby's English Botany*.

Pierandrea Mattioli (1501–77) was one of the pre-eminent herbalists of sixteenth-century Italy, and was much influenced by Dioscorides. He became physician to the Archduke Ferdinand and then the Emperor Maximilian II. His main contribution to botany was his *Commentarii in sex libros Pedacii Dioscoridis*, first published in 1544 and tinkered with for the rest of his life. In this work he describes all the plants of which he knew, including some that were previously uncatalogued.

Culpeper gives a good synthesis of all the folk treatments

Corn mint in flower.

known for spearmint and pennyroyal and backs it up with references to classical works or those based on classical works. Culpeper's analysis of the effectiveness of the herbs is still astonishingly relevant today. Scientific practice may change, technology may move on, but the way herbal remedies work remains much the same as it did 300 years ago.

NATIVE AMERICANS

Native Americans developed about 600 herbal formulas to treat disease. They used *Mentha arvensis* var. *villosa*, American mint, as a remedy for nausea and skin irritations. Like all cultures living close to nature, their knowledge of herbs and how to use them was in part built up over generations of use and in part combined with an element of magic and spirituality.

THE SHAMAN

In Native American culture, the shaman (or medicine man) is the spiritual and physical healer of the tribe, the link and guide to the spirit world, and the interpreter of dreams. He is charged in particular with preventing, diagnosing and curing illness. It is believed that sickness can be caused by evil spirits, witchcraft or breaking taboos: it is his task to restore health and adherence to the spiritual beliefs and practices. He uses herbs and sacred plants in his treatments and curing rituals, as well as the traditional sweat lodge. Magic and spirituality play a large role in his beliefs and practices.

SWEAT LODGE

A sweat lodge is a cave or circular stone structure that works rather like a sauna. Inside, a fire is lit underneath large stones, and water sprinkled on the hot stones gives off steam. Traditionally, sweat lodges have been

A Native American shaman from the Blackfoot tribe dressed in ritual attire.

used in purification rituals and to encourage dreams and visions.

An eighteenth-century herbalist called Samuel Thomson from New Hampshire approved of the sweat lodge as a technique to cure disease. He believed that all diseases came about because of the patient's exposure to cold, and he

Opposite: A mask worn by a member of the Iroquois False Face Society in a ritual to appease spirits and find cures for disease.

recommended in his *Improved System of Botanic Practice of Medicine* that sweating and steam baths be used as treatments for all ailments. This system of treatment was imported to England, where it remained popular for some time.

CURING RITES

Great ceremony – including drums, rattles, song and dance – is an integral part of treating disease and wounds in Native American culture. The Omaha Buffalo Society, an elite group of individuals to whom the buffalo spirits have appeared in dreams and visions, believe that they have special knowledge of medicine and how to cure wounds. The buffalo spirits themselves are believed to have imparted the knowledge to the individual.

The Iroquois tradition holds that spirits with ugly and deformed faces possess the power to make mankind ill. According to mythology, these spirits of disease damaged their faces when the Iroquois' creator dashed them against a mountain in his attempt to protect mankind. The members of the False Face Society don masks with deformed faces in ceremonies intended to appease the spirits and find cures for diseases.

The curing ceremonies of the False Face Society are held in January and February each year. During the ceremony a masked man dips his hands into hot ashes and places them upon the patient. He then blows through cupped hands and rubs his hands on the patient. If the spirits are appeased the patient will be cured.

INTO THE PRESENT

The herbals of the Renaissance, including that of Culpeper, were in mainstream medical use up until the development of the modern drug industry. During the scientific revolution of the nineteenth century, chemists analysed and understood the constituents of herbs and their essential oils, and were able to

introduce new drugs based on chemicals rather than natural substances. While medical chemists confirmed traditional therapeutic uses of mints and the efficacy of traditional remedies, medicine passed into the hands of the professional, and home-made herbal remedies lost their credibility.

Chambers's Encyclopaedia: A Dictionary of Universal Knowledge, published in 1908, describes the three major species of mint: spearmint, peppermint and pennyroyal:

At a market in China, a woman selling herbal medicines takes a patient's pulse.

All these species, in a wild state, grow in ditches or wet places. All of them are cultivated in gardens: and peppermint largely for medicinal use and for flavouring lozenges ... Peppermint is a powerful diffusible stomachic, and is much employed in the treatment of gastrodynia and flatulent colic. It is also extensively used in mixtures, for covering the taste of drugs. Pennyroyal and spearmint are similar in their action, but inferior for all purposes to peppermint.

The *British Herbal Pharmacopoeia* currently lists peppermint for the treatment of intestinal colic, flatulence, cold, vomiting in pregnancy and painful periods. It also recommends that pennyroyal be used for flatulent dyspepsia, intestinal colic, cold, delayed periods and gout.

At the cusp of the millennium there is a renewed desire to understand and harness the healing power of herbs in home-made recipes for remedies, cosmetics and scented decorations, and in making visits to traditional herbalists and homeopaths.

CHAPTER 2

MYTHOLOGY

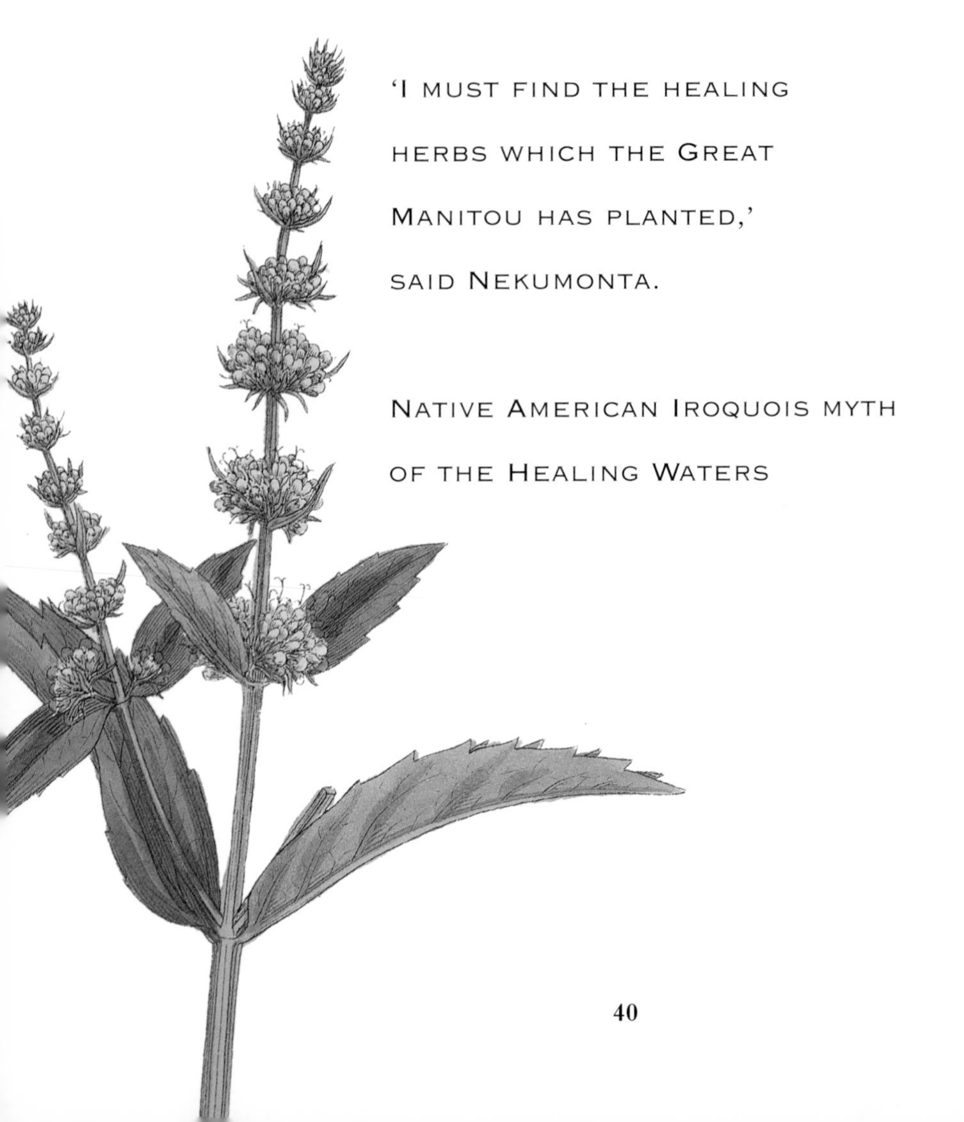

'I MUST FIND THE HEALING HERBS WHICH THE GREAT MANITOU HAS PLANTED,' SAID NEKUMONTA.

NATIVE AMERICAN IROQUOIS MYTH OF THE HEALING WATERS

ANCIENT GREECE

The mythology of ancient Greece is shot through with references to mint and other aromatic plants. Their mystical significance indicates that they played an important role in both the physical and spiritual well-being of the people.

HADES AND PERSEPHONE

According to legend, the plant mint originated as a nymph called Minthe. Hades, lord of the Underworld (called Pluto by the Romans) and brother of Zeus (also called Jupiter), spends the vast majority of his time reigning over the land of the dead. The Greeks named him Hades, which means 'invisible', because the Underworld was so dark; his nickname, 'ploutos', means 'wealth'. However, with a great lust for women, and especially nymphs, he journeys up to the land of the living from time to time.

Nymphs were divinities of nature, sometimes considered daughters of Zeus. They lived simply in the countryside where they represented the fertility, grace and spirits of nature: 'the playing of nymphs in woods and fountains', according to John Keats.

The dashing Hades with chariot and horses tries to seduce the nymph Mintha.

On one occasion, Hades glimpses the nymph Minthe and is overwhelmed by desire for her. She is greatly impressed by his golden chariot and four black horses. Hades' wife Persephone (also called Proserpine), intervenes just before Hades has the chance to seduce Minthe. In a jealous rage, Persephone metamorphoses Minthe into the plant mint. Not without justification, one would imagine, as Hades gets up to the same trick subsequently with the nymph Myrtle.

According to Robert Graves in *The Greek Myths*, mint,

rosemary and myrtle were popular herbs for use in funerary rites because they went some way to disguise the smell of putrefaction. It is perhaps owing to this that mint has become associated with the lord of the Underworld.

DEMETER

In another episode in Greek mythology, Demeter (also known as Ceres, mother of Persephone) is found at Eleusis by Abas, son of Celeus, drinking plentifully and thirstily from a jug of barley-water flavoured with mint. Demeter does not take kindly to Abas saying 'Oh, how greedily you drink!' and metamorphoses him into a lizard.

Demeter was goddess of agriculture and associated with the ritual celebration of the death and rebirth of the corn and also the celebrants' purifying death and mystical rebirth. Her links with mint are possibly explained by her associations with both agriculture (corn mint grows in corn fields in the wild) and death (mint is a funerary herb). In the earlier quotation from Culpeper on pennyroyal (see page 30), he lists barley-meal mixed with mint as a treatment for burnings. Perhaps Demeter had just burned herself.

METAMORPHOSIS

Metamorphosis is a frequent occurrence in the mythology of ancient Greece. Most often effected by one of the gods, it consists of a person being changed irrevocably into animal, vegetable or mineral form. The gods could metamorphose themselves, for their own ends – such as to impregnate mortal women, in the case of Jupiter becoming a shower of gold to mate with Danaë – and could then revert back to their usual form.

The scholar and poet Ovid (43BC–AD17) was born in Italy but visited Athens. His *Metamorphoses*, running to fifteen volumes, is one of the most influential works from the time. He describes in poetic form the metamorphoses conducted in Greek myth and makes insightful comments about psychology and symbolism.

NATIVE AMERICA

The Native Americans developed their own accounts of the origins of disease. According to the Cherokee, in ancient times all the animals, birds, fish and insects lived in harmony with each other and with people; they all communicated in the same language. However, the human race grew out of proportion with the other living things, who found themselves struggling to find enough space for themselves. Man then started to betray his friendship with the animals by inventing weapons and killing animals for food and clothing. The animals met in council to decide what to do to improve the situation. Some animals decided to protect themselves by creating diseases: the deer decided to inflict rheumatism on the human race; fish and reptiles chose to cause man to suffer from loss of appetite.

A nineteenth-century engraving of a Native American engaged in ceremonial ritual.

Throughout the developing disunity, the plants remained loyal friends of mankind. Upon hearing what the animals were inflicting, they decided to do what they could to help man. In a report entitled 'The sacred formulas of the Cherokees', published by the Bureau of American Ethnology in 1891, J. Mooney says:

> *Each tree, shrub, and herb, down even to the grasses and mosses, agreed to furnish a remedy for some of the diseases named, and each said: 'I shall appear to help man when he calls upon me in his need.' Thus did medicine originate, and the plants, every one of which has its use if we only knew it, furnish the antidote to counteract the evil wrought by the revengeful animals. When the doctor is in doubt what treatment to apply for the relief of a patient, the spirit of the plant suggests to him the proper remedy.*

CHAPTER 3

REMEDIES

TAKE PENNYROYAL, AND SEETHE IT IN RUNNING WATER, AND DRINK A GOOD DRAUGHT OF THE DECOCTION AT NIGHT GOING TO BED, WITH A LITTLE SUGAR IN IT.

NICHOLAS CULPEPER, *ENGLISH PHYSITIAN*, 1653

Left:
A Touareg man from Algeria making mint tea.

Mint is at once cooling and, because of the menthol content, particularly in peppermint, warming. Peppermint and spearmint are the two species most used but not the only ones. See pages 9–11 for details of other species and cultivars used.

The scent of mint has for centuries been considered by itself to be a general tonic for old and young alike, and to increase a person's positive outlook on life. Mint tea is valued in the hot countries of the Middle East as it gives a clean, refreshing sensation to the mouth. Mints are invigorating in baths, showers and oral preparations. They improve the digestion and help ease nausea and flatulence. This chapter gives you basic recipes that may be easily made, as well as simple remedies for a variety of common ailments.

Below:
An illustration of peppermint, from *Sowerby's English Botany*.

PARTS USED

- whole plant
- leaves
- essential oil

EFFECTS

- Analgesic: eases pain.
- Anti-emetic: helps prevent nausea and vomiting.
- Antiseptic: controls or prevents infection.

- Anti-spasmodic: reduces involuntary spasms.
- Anti-viral: stops the growth of viruses.
- Bile stimulant: stimulating a substance that aids digestion of fats.
- Carminative: helps to relieve flatulence, indigestion and colic.
- Decongestant: unblocks catarrh.
- Diaphoretic: increases perspiration.
- Digestive tonic: aids digestion.
- Peripheral vasodilator: encourages blood circulation to the surface and extremities.

CONSTITUENTS

- Volatile oil (mainly menthol in the case of peppermint; or carvone, in the case of spearmint).
- Tannins (antiseptic and anti-inflammatory).
- Flavonoids (improve circulation and strengthen blood vessels).
- Tocopherols (including vitamin E, an antioxidant and free radical binder which ameliorates the effects of the ageing process and cancer).
- Choline (part of the structure of cell membranes, acting on the nervous system).
- Bitter principles (stimulate digestion and improve liver function).

FORMS

Mint may be taken in the following forms:

- tisane
- tincture
- syrup
- lotion
- capsule

- steam inhalation
- massage rub
- compress

ESSENTIAL OIL

Essential oils are traditionally stored in blue bottles to keep out the light.

Essential oils are distilled volatile oils – chemical compounds that combine with other complex substances naturally occurring in the plant. Essential oils are a very potent and intense way of appreciating the aromatic qualities of a plant, and can reach the human bloodstream in as little as 20 minutes after application to the skin. Essential oils are used in the treatment of muscle pains, digestive disorders and inflammations. They are disinfectant and repel flies and mice.

Essential oils should not be taken internally or if you are pregnant; great care should be taken with them in the case of breastfeeding or young children (see page 59). As mint essential oils may cause an allergic reaction, always do a patch test: apply one drop of the oil mixed in a teaspoon of almond oil to a small area of moderately sensitive skin, such as the inside of the arm. Wait for half an hour to see if any allergic reaction occurs before using the oil in any other way.

A large clump of corn mint.

Peppermint and spearmint oils stimulate the digestion; peppermint oil is also antiseptic and anaesthetic to a degree. The essential oil of peppermint has a grassy and minty aroma. It may be used in conjunction with the essential oils of rosemary, lavender, lemon, eucalyptus and marjoram. Corn mint essential oil has a fresh and minty smell; however, it is better to use peppermint oil as this has a more refined aroma. Corn mint essential oil is used in cough lozenges and herb teas and syrups. It is also widely used in cosmetics, toothpastes, soaps and perfumes. The drug industry frequently uses corn mint essential oil as the base material for the production of menthol.

BASIC RECIPES

Mint is particularly versatile because it may be used to warm up or cool down the patient, as appropriate. It may be taken internally, in the form of a tisane, for example, or externally, as a massage oil or ointment. Here are thirteen basic recipes which anyone can make. You will need a small amount of equipment:

- small set of kitchen scales
- dark glass bottles
- glass jars with airtight lids
- small glass pots with airtight lids
- a fine-mesh sieve.

Always wash your hands thoroughly before starting work, and use clean and dry tools that have not been contaminated with foods or other herbs.

It is imperative to label and date any preparation that you make at home. Include the following information: the precise scientific name of the herb, the part used, the type of preparation made and the date.

NB: Store all raw ingredients and herbal preparations out of the reach of children.

INTERNAL APPLICATIONS

Here are recipes for three versatile treatments which may be taken on their own or used as the base for another preparation. For example, a tisane may be drunk or poured into a bath.

It is important not to make the recipes any stronger than is described here, not to take too much of the preparation and not to mix preparations.

In general, when water is included in a recipe, use spring water. You may use dried mint leaves or three times the amount of fresh leaves.

TISANE, TEA OR INFUSION

A tisane is a traditional medicinal preparation: indeed the word is derived from the ancient Greek word for a medicinal brew. This is a simple and popular way of enjoying the therapeutic qualities of many soft-leaved herbs such as mint.

Put 75 g (3 oz) of fresh mint leaves into a teapot. Pour on 500 ml (18 fl oz) of freshly boiled water and put the lid on the teapot to prevent the aromatic constituents from escaping. Allow the leaves to steep for 10 minutes. Then strain the liquid through a sieve straight into a cup for immediate consumption or into a jug, if you wish to drink it later. Keep the remainder in the fridge to be reheated or drunk cold as you choose during the day. Mint tea is particularly refreshing when drunk chilled.

- These quantities make three doses: a single day's supply.

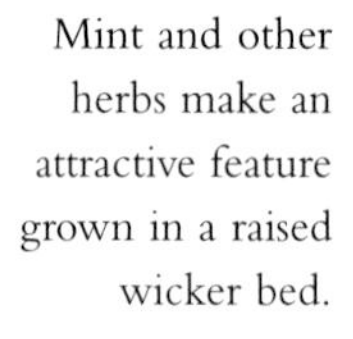

Mint and other herbs make an attractive feature grown in a raised wicker bed.

- For a single cup, use 3 teaspoons of leaves to 1 cup (225 ml/8 fl oz) of boiling water.
- Try combining mint leaves with other herbs for a fruitier tea: peppermint combines well with rosehip, dried roses, hibiscus and lemon verbena, for example.

TINCTURE

A tincture extracts the medicinal properties of a herb by using alcohol. It lasts for up to two years since the alcohol acts as a preservative. In this way, mint can be used all year round, not just when the plant is freshly harvested.

Put 900g (2 lb) of fresh mint leaves into a large jar with a tight-fitting lid. Add 700 ml (25 fl oz) of vodka and 800 ml (28 fl oz) of spring water. Put the lid on the jar and shake the jar. Store the jar in a cool, dark place for two weeks, giving it a shake occasionally. At the end of this time, strain the liquid through a sieve into a dark glass jar and store out of sunlight. Label and date the jar.

SYRUP

This is a useful recipe for treating colds and 'flu.

Pour 500 ml (18 fl oz) of boiling water over 4 tablespoons of fresh mint leaves. Cool and strain the mixture. Then place it in a saucepan with 6 tablespoons of sugar. Heat the mixture gently until all the sugar has dissolved, then boil it until it thickens.

EXTERNAL APPLICATIONS

Here are ten recipes for preparations that may be given externally. They include treatments that are rubbed in, inhaled, used for massage, and placed directly on wounds.

A mint decongestant chest rub relieves chesty colds.

DECONGESTANT CHEST RUB

For a warming chest rub, mix 1 teaspoon of almond oil with 5 drops of peppermint essential oil, 5 drops of eucalyptus oil, and 5 drops of thyme oil. Rub the mixture into the chest and back at bedtime to relieve chesty colds.

INFUSED OIL

This recipe is used as a base for making ointments or massage oils. There are two methods of infusion: hot and cold. Both have the same effect and are used in the same way, but the warm-infused oil is ready sooner.

- For the cold-infused method, fill a large jar with mint leaves; cover them with an oil rich in essential fatty acids, such as walnut, almond, olive or sunflower oil. Fit the lid, and leave the jar on a sunny windowsill for three weeks, shaking it daily. After this time, strain the oil into a clear jar, add more fresh mint, and leave the jar for a further two to three weeks. Strain the oil a second time and pour it into dark bottles with an airtight lid. Label and date the bottles.
- For the warm-infused method, place the mint leaves in a bowl and cover them with oil. Place the bowl over a saucepan containing simmering water. Heat the mixture for one hour, taking care that the saucepan does not boil dry, then strain it and heat the infused oil again with fresh mint. Strain it a second time and allow it to cool before bottling, labeling and dating it.

MINT STEAM INHALANT

This treatment is good for coughs, colds, catarrh and asthma. Be aware, however, that inhalants have high concentrations of

A mint steam inhalant can relieve the irritation caused by coughs, colds and congestion.

essential oils and should be used with caution.

Heat up 500 ml (18 fl oz) of mint tisane (see page 49) until just boiling, and pour it into a bowl. Add 5 drops of mint essential oil and 5 drops of eucalyptus essential oil and stir it in. Sit in front of the bowl, with a towel over your head and the bowl, and inhale the steam for ten minutes. Then remove the towel and sit quietly for about a further fifteen minutes before moving away. This gives the mucous membranes time to cool down gradually. For a gentler aroma use a mixture of lavender and peppermint essential oils.

Creams containing mint can soothe irritated skin.

COOL MINT CREAM

This recipe makes a gentle cream to cool and soothe itchy skin conditions. It is easily absorbed into the skin.

Combine 100 g (3½ fl oz) of almond oil or infused mint oil, 25 g (1 oz) of white beeswax (grated) and 25 g (1 oz) of anhydrous lanolin in a bowl over a saucepan containing gently simmering water. Keep the water hot until the solids melt. Add 25 ml (1 fl oz) of glycerol and 60 g (2¼ oz) of dried mint leaves, and heat the mixture for three hours (taking care that the saucepan does not boil dry). Sieve the mixture into a bowl and stir it continuously until it has cooled. After this, store the cream in airtight, dark jars in the fridge. It keeps for about two months.

MINT OINTMENT

The following recipe makes an ointment, or a base for massage oils.

Melt 25 g (1 oz) of beeswax and 25 g (1 oz) of anhydrous lanolin in a bowl over a saucepan containing gently simmering water. Add 100 ml (3½ fl oz) of infused oil (see page 52). Pour the ointment into dark glass jars and allow it to cool. Fit airtight lids on to the jars. Label and date the jars.

Essential oils are dropped into a bottle before the oil is added.

MASSAGE OIL

Place 20 drops of mint essential oil (or a combination of mint with lavender or rosemary) in a dark bottle. Add 50 ml (2 fl oz) jojoba oil. Put the lid on and then roll the bottle gently between the hands to mix the ingredients. Never shake the bottle. Label and date the bottle and store it in a cool, dark place.

Jojoba oil makes the skin feel as smooth as satin. Alternatively, you could use:

- almond oil: this is very popular in massage oils.

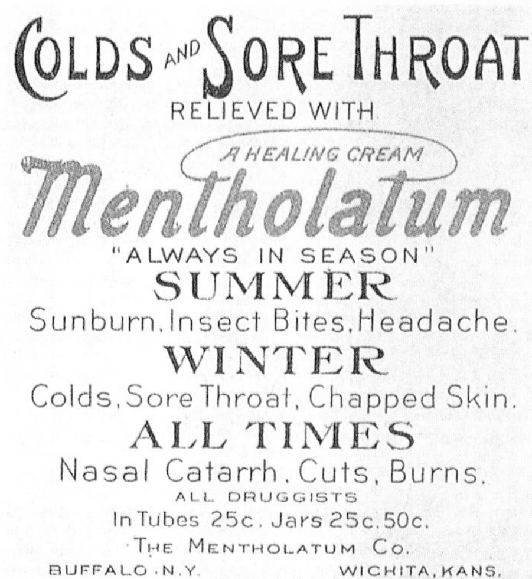

Commercial preparations containing mint can soothe a multitude of ailments.

- olive oil: calming and pungent.
- hazelnut oil: nourishing and stimulating.

MINT BATH

Add between 2 and 5 drops of mint essential oil to your bath water for a relaxing treatment for catarrh and fatigue. Eau-de-cologne mint, *M.* x *piperita* f. *citrata*, has a particularly delicious aroma. Or, instead of essential oil, you may add 500 ml (18 fl oz) of mint tisane to the same effect.

Alternatively, gather and tie together a few stems of mint and swish them through the bath.

MINT POULTICE

A mint poultice draws out infection and soothes and warms the affected part.

To make a poultice place a handful of peppermint or

spearmint leaves in a bowl of boiling water until they have softened. Allow them to cool slightly then put them between two layers of gauze and apply to the affected part.

COOL MINTY COMPRESS

For relief from inflammation, make a cool compress from peppermint or spearmint tea at normal strength or by mixing one part mint tincture with four parts water. Allow the liquid to cool then soak a clean cloth in the mixture – a swab of cotton wool for a small area or a clean flannel for a larger area – and wrap it around the part that is painful.

HOT MINT COMPRESS

For a warm treatment for sore joints, toothache or earache, fill a bowl with just-boiled water and add 4 drops of mint essential oil. Stir to mix. Soak a clean flannel in the liquid, wring it out and place it on the affected part until the liquid has cooled to blood heat. Repeat as many times as you feel necessary, or until the water in the bowl has cooled.

PREGNANCY, BREASTFEEDING AND CHILDREN

Pregnant women need to take particular care when using herbal remedies in any way: they may interfere with the actions of the uterus and sensitize the unborn child in some way (as has been discovered with peanuts). Similarly, women who are breastfeeding need to be aware of the effect a herb may have on milk supply. Babies and children have sensitive stomachs and nervous systems which can tolerate neither the intensity nor the quantity of herbs that adults can, and which are more sensitive to the toxins in herbs.

PREGNANCY

Peppermint and spearmint may be used to alleviate sickness in pregnancy. Try taking a cup of mint tea three times a day. Commercial tisane preparations are generally considered safe in pregnancy, but be aware that if taken in excess they may do more harm than good. Pennyroyal, *Mentha pulegium*, is not suitable for use in pregnancy, however, as it stimulates the uterus and may cause abortion.

BREASTFEEDING

It has traditionally been believed – and documented – that mint (especially corn mint) can lower the milk supply of a breastfeeding mother. However, if breastfeeding has established itself and is proceeding satisfactorily small doses of mint on occasion, as part of a varied diet, seem to have no recorded detrimental effect on milk supply. The La Leche League, the international charity that offers mothers breastfeeding help and information, stresses, however, that all herbs should be considered as medication, and that if any difficulties arise with breastfeeding, the mother should seek help and declare that she has been taking mint.

CHILDREN

Do not give the herb peppermint in any form to children of less than four years, or any mint essential oils to children younger than ten. Catmint, *Nepeta cataria*, may be used as an adequate and safe substitute with children younger than this. Spearmint is milder than peppermint and is generally a better choice for children.

If symptoms of illness persist or worsen, consult a professional immediately. Take particular care not to overdose your child: see the recommended doses for children below. Take care not to give your child several different treatments at the same time.

Colic

An older baby or child may be given a weak spearmint tea to drink (see Recommended Doses for Children, opposite) if they have colic or a sore tummy, or are teething. Alternatively, give an older child a massage that includes spearmint essential oil (see page 54). **Do not use peppermint essential oil.**

Coughs

In the case of a cough, spearmint may be combined with black horehound (*Ballota nigra*). This is an expectorant herb which calms spasms and eases feelings of nausea. Black horehound stimulates the uterus, however, and so should not be used by pregnant women for their own treatment.

Diarrhoea

Try giving your older child a gentle massage with 1 drop each of spearmint and lavender essential oils mixed in a little warm almond oil.

Earache

For children over ten *only*, mix up 1 drop of spearmint essential oil in 1 tablespoon of warmed almond oil. Soak a cotton wool ball in the mixture and insert it gently into the ear. Do not poke it in lest you damage the ear. For children between four and ten use either catmint essential oil or warm peppermint tea. If the pain continues, seek medical advice.

Fever

A poorly child with a fever can be comforted with a few teaspoons of mint tea, and by having their face and hands wiped with cooled mint tea.

Sleeplessness

A way to give relaxation and pleasure to both mother and older child is to massage the child with a mixture of 20 ml (¾ fl oz) sweet almond oil and 1 drop of mint essential oil. Alternatively, for children younger than ten, use catmint. A weak spearmint tea, perhaps flavoured with a little honey, can also help at bedtime.

RECOMMENDED DOSES FOR CHILDREN

NB DO NOT GIVE PEPPERMINT IN ANY FORM TO CHILDREN UNDER FOUR YEARS, OR ANY MINT ESSENTIAL OIL TO CHILDREN YOUNGER THAN TEN.

Age	Dose
6 months–1 year	5% of adult dose
1–3	10%
3–5	20%
5–7	30%
7–9	40%
9–11	50%
11–13	60%
13–15	80%
15 upwards	100%

A fifteenth-century French illustration showing a garden of medicinal plants.

GENERAL AILMENTS

Mint has been used successfully over thousands of years to treat a wide range of complaints including aches and pains, indigestion and liver problems, nausea, period problems, fevers, coughs, colds, catarrh, sinusitis, itchy skin conditions, as well as to freshen the breath and clean the teeth. For major or persistent problems be sure to consult a qualified herbalist, homeopath, or your medical practitioner.

Acne

To open and cleanse the pores, take a steam inhalation with peppermint or spearmint essential oils (see page 52).

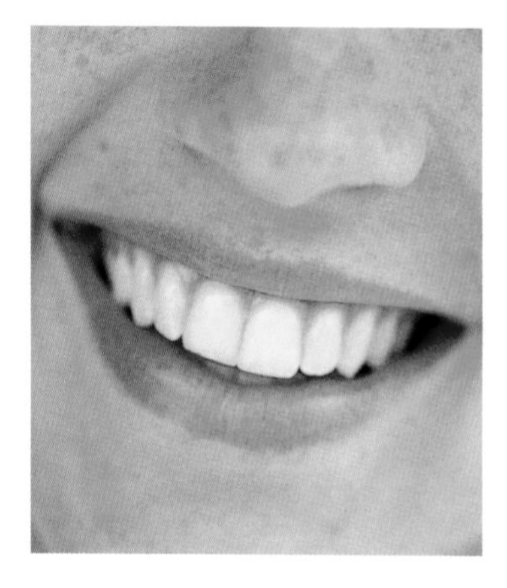

Peppermint is effective for freshening the teeth and mouth.

Asthma

In the case of mild asthma attacks take a peppermint or spearmint steam inhalation (see page 52). The antispasmodic action of the mint together with the menthol (particularly with peppermint) provide relief. If you are suffering a more major attack, consult your doctor at once.

Bad breath

Peppermint is one of the most effective and widely used treatments for bad breath as it has a role to play in freshening the mouth and teeth. In general, however, try not to eat a great number of peppermint sweets in the hope of sweetening the breath as the sugar can cause as many problems as the peppermint solves, and the peppermint itself may irritate the stomach. Instead, drink a cup of peppermint tea.

Bronchitis

To ease bronchitis, have a massage with peppermint or spearmint essential oil or prepare a steam inhalation.

Burns and scalds

To soothe burns, soak a pad of cotton wool in cool mint tea and place it over the affected part. Alternatively, mix 2 drops of peppermint essential oil, or one of peppermint and one of lavender, in 1 tablespoon of almond oil to calm the heat of a burn. For more severe burns, consult your doctor.

Catarrh

To relieve catarrh and a painful chest, mix together 5 drops of peppermint essential oil and 5 drops of eucalyptus essential oil with 20 ml (¾ fl oz) of almond oil. Rub onto the chest. A steam inhalation is also a good treatment (see page 52). Make a tisane containing equal parts of peppermint, elderflower and yarrow to relieve catarrh.

Colds and 'flu

Make a steam inhalant of peppermint, sage and lavender to bring gentle relief from the symptoms of a cold or 'flu.

Coughs

Have a massage or steam inhalation with peppermint or spearmint essential oils.

Cramp in the stomach

To ease the pain of gastric cramp, apply a hot compress or have a peppermint or spearmint stomach massage.

Cramp

Mix 2 drops of pennyroyal essential oil into 1 tablespoon of almond oil and rub it into the affected area. Alternatively, an old remedy from the Fens area of England recommends that the fresh leaves of pennyroyal be crushed and rubbed into the affected part.

Diarrhoea

To calm the stomach, drink a cup of warm mint tea or apply a warm peppermint compress.

Dizziness and fainting

Mints are invigorating and therefore make a good, simple treatment for dizziness. Wrap a handful of fresh leaves in a handkerchief and waft them under the nose of the person who has fainted. Then try a gently reviving warm tisane of peppermint, perhaps sweetened with a little honey.

Mint has long been used in the treatment of illness.

Earache

For warming relief from earache, mix 1 drop of peppermint essential oil to 1 tablespoon of warmed almond oil. Soak a cotton wool ball in the mixture and place it gently but firmly into the ear. Do not push it into the ear canal. If symptoms do not improve, seek advice from your health professional.

Fevers

Drink a cup of mint tea or apply a cool compress. To make the patient feel more comfortable wipe the face and hands with cooled mint tea.

Flatulence

Take mint tea up to three times a day to ease flatulence. Alternatively, massage the stomach and abdomen with mint essential oil.

Hangover

Traditional remedies suggest drinking barley-water with mint, or peppermint tea with a splash of lemon juice.

Headache accompanying fever
Have a mint steam inhalation or apply a cool compress to the forehead to ease a feverish headache.

Hiatus hernia
A tisane made from peppermint eases the symptoms of hiatus hernia.

Indigestion
After meals, drink a cup of mint tea, or a combination of mint and meadowsweet tea, to calm most digestive troubles. Alternatively, try taking 1 teaspoon of peppermint syrup.

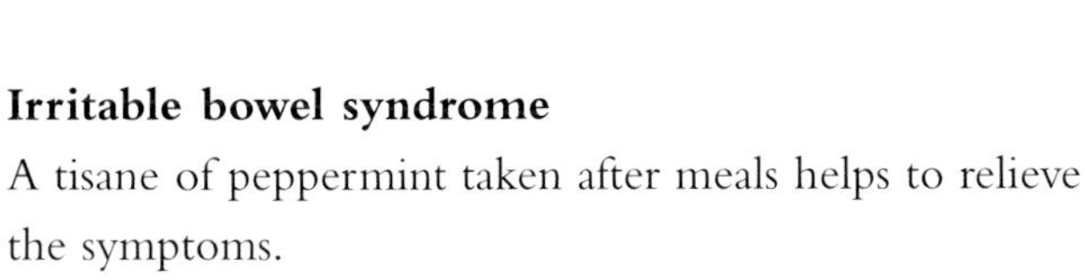

Irritable bowel syndrome
A tisane of peppermint taken after meals helps to relieve the symptoms.

An illustration of wild peppermint, from *Sowerby's English Botany*.

Laryngitis and sore throats
There are two effective treatments for laryngitis and sore throats. Make up a gargle with either mint tea or one part mint tincture to five parts water. Alternatively, use a peppermint steam inhalation.

Migraine
In the case of a migraine or sick headache, apply a cool compress or a poultice of crushed peppermint and lavender leaves to the head. Alternatively, take a tisane made from mint leaves and either lavender or camomile flowers in equal proportions to ease nausea.

Motion sickness

Travel sickness may be effectively calmed by simply holding under the nose a handful of mint leaves wrapped in a handkerchief. Be careful if using this treatment with children: spearmint is generally better for them as it is less strong. Do not overdo it, or any benefit will be negated.

Nausea

Have a cup of mint tea or a dessert spoonful of tincture three times a day to combat all forms of nausea, including travel sickness and morning sickness. If you are pregnant, see Pregnancy, breastfeeding and children, page 56.

Nervous exhaustion

A rejuvenating mint bath helps to bring relief to someone suffering from nervous exhaustion or tension.

Neuralgia

In the case of neuralgia, a massage with diluted mint essential oil, or a bath scented with mint essential oil, brings relief.

Period pain

Drinking mint tea, applying a hot mint compress or taking a massage will soothe some of the cramps associated with period pain. Pennyroyal is particularly suitable in the case of light but painful periods.

Poor concentration

If you are not able to concentrate, or not feeling very alert, sniffing a handful of fresh peppermint leaves – or a drop of essential oil rubbed into the back of the hand – will revivify you.

Rheumatism

Tiger Balm is a classic proprietary remedy for rheumatism. It consists of a blend of menthol, peppermint, camphor, cajuput, cassia and cloves. More simply, try mixing 2 drops of peppermint essential oil in 1 tablespoon of warmed almond oil, and rub it into the aching joint.

Sinusitis

A peppermint steam inhalation with added essential oil of eucalyptus helps to ease the pressure and pain of sinusitis. Peppermint and ginger tincture, combined in equal parts, is also a good treatment.

Skin irritations

For hot and itchy skin conditions apply cooling mint cream (see page 53).

Sleeplessness

Drinking a warm cup of mint tea at bedtime helps induce restful sleep. In the case of children, see page 59.

Travel sickness

For relief from moderate symptoms, have a cup of mint tea before travelling. In the case of children, see the box on children's dosages, page 59.

CHAPTER 4

COSMETICS

She rose where she had falled down, and called her maid, and went down into the house, in which she abode in the sabbath days, and in her feast days, and pulled off the sackcloth which she had on, and put off the garments of her widowhood, and washed her body all over with water, and anointed herself with precious ointment, and braided the hair of her head, and put on a mitre upon it, and put on her garments of gladness, wherewith she was clad during the life of Manasses her husband. And she took sandals upon her feet, and put about her her bracelets, and her chains, and her rings, and her earrings, and all her ornaments, and decked herself bravely, to allure the eyes of all men that should see her.

***Judith* X:2–4**

Everyone knows that mint is traditionally used in oral preparations 'for fresh, minty breath', as the advertising slogans usually run. It does, however, have considerably more cosmetic uses than just this. It generally refreshes, energizes and revitalizes both physically and mentally; literally from the head down to the feet.

A fresco of cupids making cosmetics and perfumes, from the House of the Vettii, Pompeii, first century AD.

MAKING YOUR OWN COSMETICS

There follows a collection of simple recipes for you to make your own cosmetics. Try the ones that appeal, and then don't be afraid to experiment. Mint combines well with lavender and rosemary, and also with grapefruit essential oils for extra zest.

Be aware that sensitive skins may have an adverse allergic reaction, so always do a patch test (see page 48) before using your home-made cosmetic.

Opposite: A perfumed oil container in the form of a siren, made by the Milesians (Greeks) in c. 525BC.

TEETH AND MOUTH

Peppermint and spearmint have for generations been used as a way of freshening the teeth and mouth, and continue to be a recipe for success in commercial mouthwashes and toothpastes.

MINT TOOTHPASTE

For a day's supply of refreshing toothpaste, mix 2 teaspoons of sodium bicarbonate with 2 teaspoons of vegetable glycerine. Then add 6 drops of either peppermint or spearmint essential oil. Peppermint has a stronger taste; spearmint is a more gentle taste and is more suitable for children. Brush the teeth as normal. Seal the remainder of the toothpaste in a small jar with an airtight lid.

A mint foot bath cools and refreshes tired feet.

MINTY MOUTHWASH

A minty mouthwash is particularly refreshing. Steep 4 tablespoons of fresh mint leaves in 1 litre (1¾ pints) of boiling water. Let it cool, then refrigerate it until it is chilled. After this, strain the mixture and bottle it.

FEET

Mint restores and soothes tired, hot feet. Try the following foot recipes.

MINT FOOT BATH

For a cooling foot bath, half-fill a basin with cool water and add 2 drops of mint essential oil. Then sit back and relax.

SOOTHING FOOT CREAM

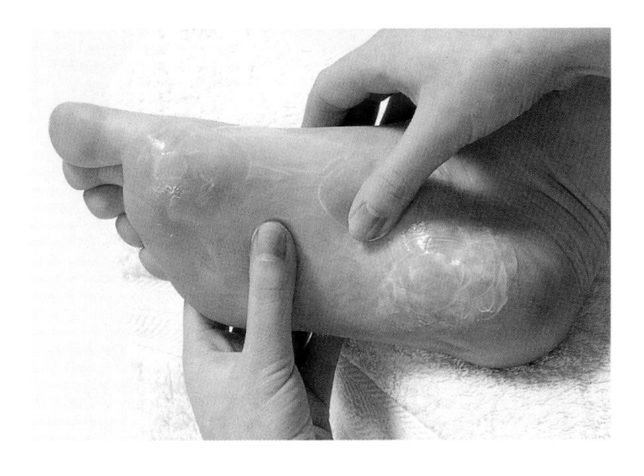

Treat sore feet to a soothing, minty foot cream.

This recipe makes a soothing and fragrant foot cream. Heat in a saucepan 30 g (1 oz) of cocoa butter; 30 g (1 oz) of grated beeswax; 135 ml (5 fl oz) of almond oil; warm up 90 ml (3 fl oz) mint tea and 1½ teaspoons of borax; add the two together and stir; add 30 drops of mint essential oil (or 15 of mint and 15 of lavender) when cool. This keeps for up to two months in the fridge in an airtight jar. Label and date the jar.

COOLING FOOT POWDER

Mix together 1 tablespoon of talcum powder with 1 tablespoon of arrowroot and 4 drops of mint essential oil.

BATH AND SHOWER

You may use mint in various ways to enjoy a particularly refreshing and energizing bath or shower.

PEPPERMINT BATH FIZZ

For a cooling and energizing bath fizz try the following recipe. Mix together 80 g (3 oz) of sodium bicarbonate with 1 tablespoon of citric acid. Then sprinkle in 10 drops of mint essential oil (or 5 of mint and 5 of lavender). Mix together. You may make bath bombs by pressing the mixture into an egg cup until it holds together; alternatively, sprinkle a handful directly into the bath.

MINT SHOWER GEL

For a refreshing bite to your morning shower, add a few drops of mint essential oil to unperfumed shower gel at the rate of 1 drop per 1 ml.

A minty mask will help to freshen your complexion.

FACE AND BODY

To treat your face and body, try some of the following recipes which cleanse and brighten the skin.

MINTY MASK

For a quick treatment to freshen your complexion try the following. Liquidize 2 tablespoons of fresh peppermint or spearmint leaves in 2 tablespoons of spring water. Drain off the excess moisture, so that you are left with a moist lump of mint. Clean your face as usual, then apply the mask and leave it for about 15 minutes before rinsing it off.

MINT TINGLE WASH

For a cooling wash, add a few drops of mint essential oil to a basinful of water and wash face and hands as normal.

Right: Add mint essential oil to water for a refreshing face wash.

WASH FOR ITCHY SKIN

For itchy, hot and sensitive skin, wash the face in a cool mint tea. Creams made with mint essential oil also calm the skin (see page 53).

EYE COOLER

For tired and swollen eyes make a cup of mint tea from teabags; allow the teabags to cool, then place them over your eyes.

EYE BRIGHTENER

To brighten the eyes and reduce dark circles, crush some fresh mint leaves and with your fingertips apply a small amount of the juice under the eyes. Make sure that you do a patch test first (see page 48).

BLACKHEAD REMOVER

Steam the face with peppermint essential oil mixed in a bowl of just-boiled water. After this, once the pores have expanded, use cotton wool buds gently to squeeze out the blackheads. To finish, apply the minty mask or mint tingle wash (see page 70).

MINT SPRITZER

For a refreshing spritzer for the face or body, make a tisane of mint and parsley, allow it to cool, then pour it into an atomizer and spritz it.

DEODORANT

Use a sweet-smelling mint such as eau-de-Cologne mint, *Mentha* x *piperita* f. *citrata*, or a combination of spearmint and rosemary. Put a handful of fresh herbs into a saucepan and cover them with water. Simmer the herbs for five minutes then pour the mixture into a bowl or jug and allow the liquid to cool. Strain the liquid into an atomizer and use it to spray the body. In hot weather, store the deodorant in the fridge to really energize and refresh you!

GREAT GRANDFATHER'S AFTERSHAVE

Here is an old recipe for a delightfully aromatic aftershave, devised by my grandfather, a pharmacist from the north of England.

Dissolve 2 g of menthol (available from a pharmacist) and 10 drops each of the essential oils of pine, bergamot, nutmeg, lavender and geranium in 250 ml (9 fl oz) of surgical spirit. Add 60 ml (2 fl oz) of witchhazel and a few drops of green food colouring to suit yourself. Then top up the mixture with a further 250 ml (9 fl oz) of surgical spirit. Stir well. The menthol and essential oils should dissolve completely in the surgical spirit, but in case there is any sediment, filter the mixture as follows. Take a funnel and place a ball of cotton wool in the bottom. Sprinkle a little unperfumed talcum powder on top of the cotton wool and pour the mixture through the funnel to trap any sediment. Store the aftershave in airtight bottles, labelled and dated.

HAIR

Mint is traditionally used to regulate oily hair conditions.

HAIR TONIC

Blend together 1 teaspoon of jojoba oil with 1 teaspoon of rum and 3 drops of mint essential oil. Massage it into the hair and scalp and leave it for one hour. Then wash it out with shampoo as normal.

Mint is a traditional ingredient in treatments for oily hair.

SHAMPOO

To make mint shampoo mix 4 drops of mint oil with 1 tablespoon of unscented baby shampoo. Wash your hair as normal.

CHAPTER 5

SCENTED GIFTS

BE CAREFUL, ERE YE ENTER IN, TO FILL YOUR
BASKETS HIGH WITH FENNEL GREEN,
AND BALM, AND GOLDEN
PINES, SAVORY,
LATTER-MINT, AND
COLUMBINES, COOL
PARSLEY, BASIL SWEET,
AND SUNNY THYME; YEA,
EVERY FLOWER AND LEAF OF EVERY CLIME,
ALL GATHER'D IN THE DEWY MORNING ...

JOHN KEATS, *ENDYMION*, BOOK IV

For thousands of years, mint has been brought indoors to perfume the home and repel moths, ants, mice and rats. In hot climates the cooling aroma of mint will have pervaded steamy interiors and generated a feeling of calm. This chapter shows you how to make your own scented gifts and decorations: pot-pourri, aromatic cushions, room sprays and scented candles.

MAKING YOUR OWN SCENTED ITEMS

You can make your own scented decorations quite simply. They are a pleasure to make, either for yourself or as gifts, and you can use your own creativity in choosing fabrics, ribbons, lace or any other trimmings you like. You need not do a great deal of sewing. If you do get out your sewing machine, ensure that you leave one edge of the fabric easy to undo so that you can open it up to revitalize the mint leaves with a drop of essential oil if necessary at some point in the future. Pot-pourri usually needs pepping up every couple of months.

Opposite: Selsley Herb Garden in Gloucestershire.

SCENTED SACHETS

To make scented clothes sachets to hang in a wardrobe or put in a drawer, place dried peppermint or spearmint leaves in a square of fabric of your choice. Gather together the corners and tie them with attractive ribbon or lace. Leave a loop by which to hang it if you choose to place it over a clothes hanger in the wardrobe.

Left: A pretty, scented sachet tied to the bedpost releases a fresh perfume.

FRAGRANT MINT SPRAY

For a fragrant room spray pour 500 ml (18 fl oz) of boiling water onto 12 tablespoons of fresh mint leaves, or a combination of mint and lavender. Cool and strain the mixture. Then mix it with 150 ml (5 fl oz) of vodka. Add 10 drops of mint essential oil. Store it in a dark glass bottle, clearly labeled and dated.

POT POURRI

A refreshing pot-pourri decorates and perfumes a room.

You can make a refreshing pot-pourri of scented leaves and petals incorporating mints. An enticing blend includes rose petals, peony petals and mint. For every 4 cups (900 ml/ 32 fl oz) volume of leaves and petals, add 1 tablespoon of orris root to fix the perfume, and 5 drops of peppermint essential oil. Store the pot-pourri in a sealed container in a warm, dark place for six weeks, shaking it occasionally. You may top up the essential oils after the pot-pourri has been displayed for a while, if you need to reinvigorate it.

SCENTED CUSHION

Make a cushion cover out of the fabric of your choice, leaving one side unstitched. Half fill the cushion with kapok filling. Then add a bag of mint pot-pourri (see above), add more kapok filling, and sew shut the last side.

Alternatively, a simpler approach is to make a small muslin square, fill it with mint pot-pourri and tuck it inside your cushion covers. Every time you lean against the cushions you will release an invigorating scent.

BATH SCENTER

Gather mint leaves into a square of muslin. Tie the muslin with a ribbon and loop it over the hot tap so that the warm water runs through the bag.

SCENTED CANDLE

A scented candle may be made in several ways. Most simply, light a wide candle and let the top of the wax melt. Put out the flame and mix a few drops of peppermint essential oil into the melted wax. The fragrance should last for an hour.

Alternatively, you may buy a candle-making kit and add peppermint essential oil into the mixture just before pouring it into the candle mould.

Mint-scented candles add a touch of sophistication to any room.

FURTHER READING

Arber, A., *Herbals: Their Origin and Evolution; A Chapter in the History* of Botany, 1470–1670, Cambridge University Press, Cambridge, 1986

Bown, D., *The Royal Horticultural Society Encyclopedia of Herbs and Their Uses*, Dorling Kindersley, London, 1995

Bremness, L., *Crabtree & Evelyn Fragrant Herbal: Enhancing Your Life with Aromatic Herbs and Essential Oils*, Quadrille Publishing, London, 1998

Chambers' Enclcyopedia, Chambers, London, 1908

Culpeper, N., *Culpeper's Complete Herbal & English Physician*, Parkgate Books, London, 1987

Gao, Duo, *The Encyclopedia of Chinese Medicine*, Sevenoaks, London, 1997

Graves, Robert, *The Greek Myths*, Complete Edition, Penguin, London, 1955

Hillier, M., *The Little Scented Library: Sachets and Cushions*, Dorling Kindersley, London, 1992

Lawless, J., *The Illustrated Encyclopedia of Essential Oils: The Complete Guide to The Use of Oils in Aromatherapy and Herbalism*, Element, Shaftesbury, 1995

Mooney, J.,"The sacred formulas of the Cherokees", Annual Report of the Bureau of American Ethnology (1885–86), Washington DC, 1891

Nice, J., *Herbal Remedies for Healing: A Complete A–Z of Ailments and Treatments*, Piatkus, London, 1990

Ody, P., *Simple Healing with Herbs: Herbal Treatments for More Than 100 Common Ailments*, Hamlyn, London, 1999

Shaw, N., *Herbalism: An Illustrated Guide*, Element, Shaftesbury, 1998

Spence, *The Myths of the North American Indians*, Dover, New York, 1989

Taylor, Colin F., *Native American Life: The Family, the Hunt, Pastimes and Ceremonies*, Salamander, London, 1996

USEFUL ADDRESSES

G Baldwin and Co – supplier of herbs and essential oils by post
173 Walworth Road
London, SE17 1RW
Tel: 0171 703 5550

Neals Yard has a mail-order service:
Tel: 0171 627 1949

National Institute of Medical Herbalists
56 Longbrook Street
Exeter, EX4 6AH
Tel: 01392 42602230

La Leche League
PO Box 29
West Bridgeford, Notts
NG2 7NP

ACKNOWLEDGEMENTS

Thanks go to Deni Bown, for her comments and expertise, to Frances Vargo and Margot Richardsonfor their invaluable contributions to the book, and to my father, Brian Ferry, as well as Rachel al-Azzawi and Nina Payne for their help and encouragement. Particular thanks are due to Richard, who provided me with great support and opportunities to write in such as way that Matilda didn't realize Mummy was busy.

Thanks to Collins & Brown for permission to reproduce text from Culpeper's Herbal.

The publishers would like to thank the following sources for their kind permission to reproduce the pictures in this book:

Ancient Art & Architecture Collection Ltd 28, 67; **Gillian Beckett** 8br, 9tl, 11br,11tl; **The Bridgeman Art Library**, London, **Bibliotheque Nationale**, Paris, France, Fr 9136 Frontispiece to "The Garden of Medicinal Plant", French manuscript, 15th century Garden of Medicinal Plant French (15th) 59, Eton College, Windsor UK, The Art of the Apothecary, Latin (manuscript) Apuleius Dioscorides 19, Greek Museum, University of Newcastle Upon Tyne UK, Perfumed oil container in the form of siren, Milesian, c.525 BC 66; Stapleton Collection UK, Armenian Ladies at Home, engraved by Charles Parsons, 1862 (litho) by H.J.Van Lennep (19th century) (after) 27tl; **©Carlton Books Ltd.** 44, 47, 53tl, 53br, 54, 69, 70b, 73, 75, 76, 77; **Jean Loup Charmet** 20, 21, 41, 43, 62; **Corbis**/Eric Crichton 12, Kelly-Mooney 15, Keren Su 22, Michael S.Yamashita 38; **ET Archive** 23; **Mary Evans Picture Library** 18, 25, 29tr, 32, 35, 45br, 63; **Werner Forman Archive** 37; **Garden Matters** 7; **John Glover** 8tl, 9br , 17, 29bl, 50, 74; **Frank Lane Picture Agency**/Rolf Bender 33, Eric & David Hosking 34, W J Howes 10tl, Ian Rose 10br, 40, 48; **Pictor Uniphoto** 51; **Harry Smith Horticultural Photographic Collection** 31; **Tony Stone Images**/Chris Bayley 6, Grilly Bernard 45tl, Dan Bosler 70t, Chris Craymer 68, Jerome Tisne 60; **Vintage Magazine Archive Ltd** 55.

INDEX